discovermore

Space Science

# Exploring Moons

Elizabeth Pruet

Britannica
Educational Publishing

IN ASSOCIATION WITH

THE ROSEN PUBLISHING GROUP

Published in 2026 by Britannica Educational Publishing (a trademark of Encyclopædia Britannica, Inc.) in association with The Rosen Publishing Group, Inc.
2544 Clinton Street, Buffalo, NY 14224



Distributed exclusively by Rosen Publishing.
To see additional Britannica Educational Publishing titles, go to rosenpublishing.com.



Editor: Danielle Haynes
Book Design: Michael Flynn

Photo Credits: Cover Alones/Shutterstock.com; (series background) Dai Yim/Shutterstock.com; p. 4 AstroStar/Shutterstock.com; p. 5 Evgeniyqw/Shutterstock.com; p. 6 Artsiom P/Shutterstock.com; p. 7 https://commons.wikimedia.org/wiki/File:Galileo_galilei,_telescopi_del_1609-10_ca..JPG; p. 8 DIAMOND HEART/Shutterstock.com; p. 9 aeonWAVE/Shutterstock.com; p. 10 https://commons.wikimedia.org/wiki/File:Moon_worldwind.jpg; pp. 11, 13, 14, 15, 17, 19, 21 (bottom), 22, 25, 26, 29 courtesy of NASA; p. 12 https://commons.wikimedia.org/wiki/File:Massive_Smash-Up_at_Vega.jpg; p. 16 muratart/Shutterstock.com; p. 21 (top) courtesy of European Southern Observatory; p. 23 https://commons.wikimedia.org/wiki/File:Orbits_of_inner_moons_of_Neptune_including_S_2004_N_1.jpg; p. 24 Siberian Art; p. 27 joshimerbin/Shutterstock.com; p. 28 AstroStar/Shutterstock.com.

Library of Congress Cataloging-in-Publication Data

Names: Pruet, Elizabeth author
Title: Exploring moons / Elizabeth Pruet.
Description: [Buffalo] : Britannica Educational Publishing, [2025] |
Series: Discover more: space science | Includes bibliographical references and index.
Identifiers: LCCN 2025036105 | ISBN 9781641905305 library binding | ISBN 9781641905299 paperback | ISBN 9781641905312 ebook
Subjects: LCSH: Satellites--Juvenile literature | Moon--Juvenile literature | Astronomy--Juvenile literature
Classification: LCC QB401.5 .P78 2025 | DDC 523.3--dc23/eng/20250926
LC record available at https://lccn.loc.gov/2025036105

Manufactured in the United States of America



CPSIA Compliance Information: Batch #CWBRIT26. For further information contact Rosen Publishing at 1-800-237-9932.

# Contents

# Shining Bright

Take a look up in the night sky. With no clouds, you can see many bright stars. The brightest object in the night sky is not a star. It is Earth's moon.

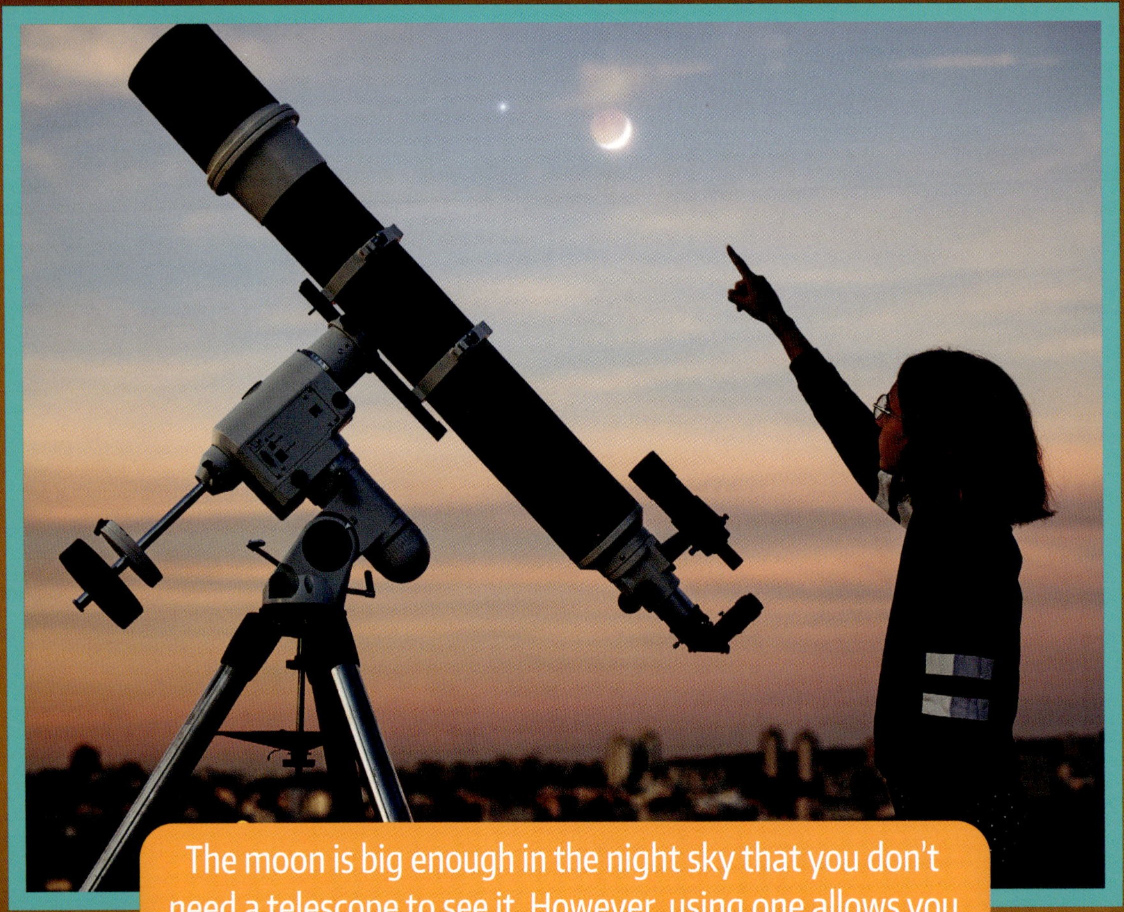

The moon is big enough in the night sky that you don't need a telescope to see it. However, using one allows you to see some of the features of the moon better.

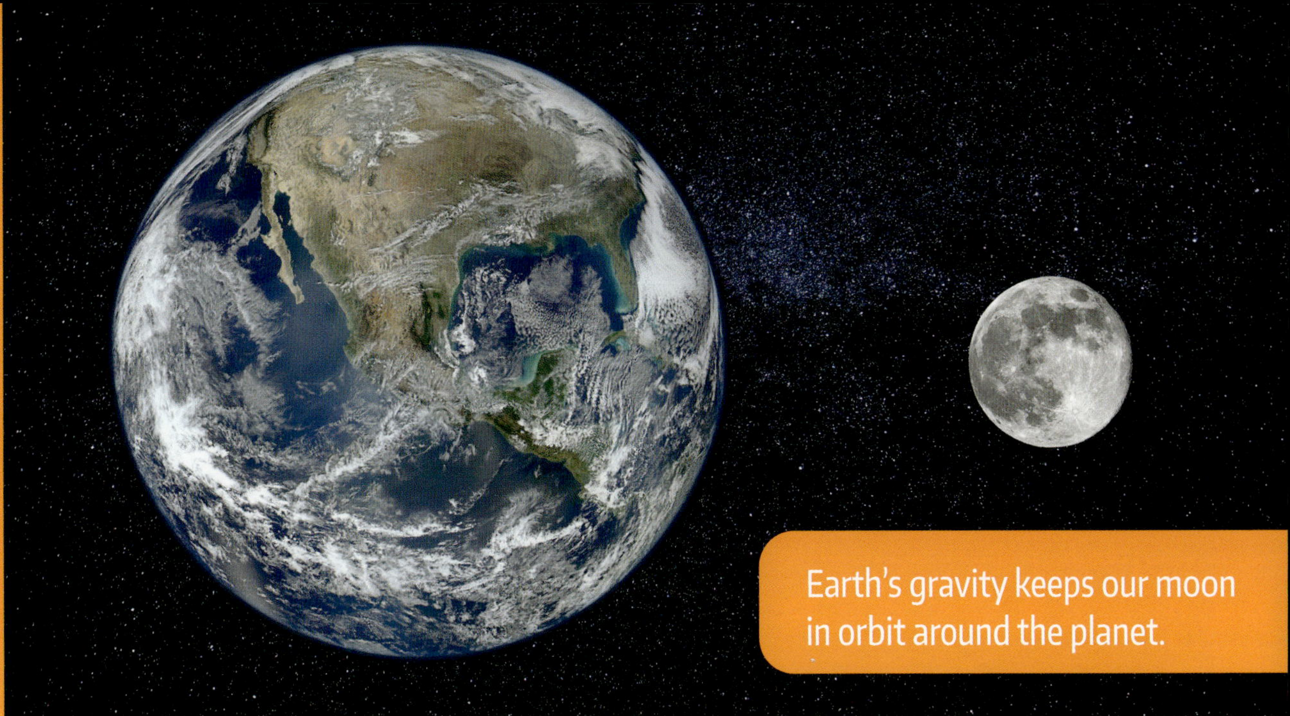

Earth's gravity keeps our moon in orbit around the planet.

A moon is a satellite, which is an object that orbits a larger object. Moons orbit planets and other large bodies because of gravity. Gravity is a force that attracts objects to the mass of other things. Planets orbit stars, such as the sun, but moons are caught in the gravity of planets.

Every planet in Earth's solar system except Mercury and Venus has at least one moon that scientists know about. There are hundreds of moons in our solar system!

## Consider This

New moons are always being discovered by scientists. Why do you think Mercury and Venus do not have any moons?

# Satellites

You may hear the term "satellite" and think of the many machines orbiting the Earth today. Humans made all the satellites around Earth except one. They put these machines into space for various reasons. Earth has only one natural satellite: the moon.

There are more than 11,000 man-made satellites orbiting Earth, the largest of which is the International Space Station.

These are believed to be the oldest surviving telescopes, from the early 1600s.

The ancient Greeks called Earth's moon a "planet." To them a planet was anything big and bright that moved in the sky. This belief meant that a star such as the sun was a planet, but Earth was not.

In the 1600s, the invention of the telescope let people observe the moon and the rest of space more carefully. Now people know there are many moons in the solar system, and there may be many more throughout the universe. Italian astronomer Galileo Galilei was the first to observe a moon around another planet in 1610—Ganymede, which orbits Jupiter.

## Consider This

The International Space Station (ISS) is a satellite that orbits Earth. What is the difference between the ISS and the moon?

# Our Moon

From what we can see on Earth, the moon looks pretty small, but it's much bigger than the dwarf planet Pluto. The diameter of the moon is a little more than one-quarter the size of the diameter of Earth. Its diameter, or distance through its center, is about 2,200 miles (3,500 kilometers). The moon is made mostly of rock. On average, it orbits about 238,900 miles (384,400 km) from Earth.

This image shows how big the moon is compared to the Earth.

During a **total solar eclipse**, all you can see on Earth is a thin ring of light around where the sun should be. It gets dark in the middle of the day on Earth!

Like the planets, the moon has two types of movement: orbit and spin. It takes the moon about 27 days to make one trip around Earth. The moon also spins about its center. It completes one rotation in about 27 days. Because the moon takes about the same amount of time to complete one orbit and one rotation, the same side of the moon always faces Earth. The moon does not give off light. It reflects light coming from the sun. Sometimes, during a solar eclipse, the moon blocks some or all of the sun!

## WORD WISE

**AN EVENT CALLED A TOTAL SOLAR ECLIPSE OCCURS WHEN THE MOON PASSES BETWEEN EARTH AND THE SUN. THE MOON BLOCKS ALL BUT THE OUTLINE OF THE SUN FOR A SHORT TIME.**

# Features of the Moon

Some people say they see the face of a man in the moon. Its surface has thousands of pits called craters that make up these imaginary facial features. The craters form when chunks of rock and metal called meteorites crash into the moon. These crashes have covered the moon's surface with rocks and dust. The moon also has plains made of lava that erupted from volcanoes billions of years ago. It has its own gravity, but it is less than Earth's.

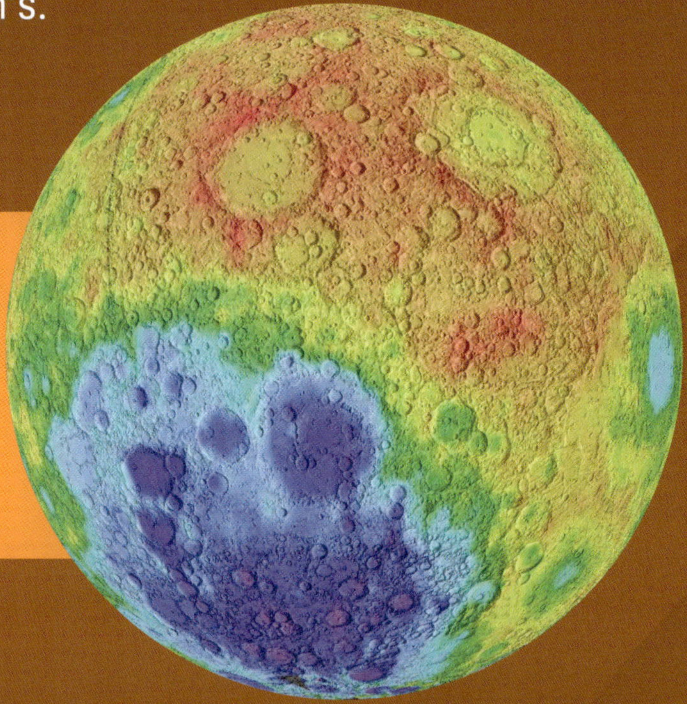

The largest crater on the moon is 1,553 miles (2,500 km) across! It can be seen in blue and purple here on this **topographic map** of the moon.

The atmosphere on Earth scatters the sun's rays and appears blue to us during the day. The moon doesn't have an atmosphere, so you have a clear view of outer space.

Meteorites have changed the face of the moon because it does not have an atmosphere to protect it. The absence of an atmosphere also leads to a wide range of temperatures on the moon. During the daytime, when the moon faces the sun, its surface temperature is about 225° F (107° C). At night, the temperature drops to about −243° F (−153° C). Without an atmosphere, the sky is always dark on the moon, even during the day.

## WORD WISE

**A TOPOGRAPHIC MAP OF AN AREA SHOWS THE PHYSICAL FEATURES AND ELEVATIONS OF A PIECE OF LAND.**

# Forming Earth's Moon

Not all moons are formed the same way. Scientists are not entirely sure how Earth's moon was formed, but they do have ideas. The leading idea is that a large rock crashed into Earth. This rock broke off part of the planet, which became the moon.

Some scientists believe a large object called Theia crashed into Earth billions of years ago. This shot material out into space that later became the moon.

Space suits help astronauts, such as Buzz Aldrin, seen here in 1969, breathe. They also protect them from extreme temperatures on the moon.

Another idea is that both Earth and the moon were formed together around the same time. Some scientists believe the moon was formed somewhere else and later got pulled in by Earth's gravity. However, these ideas can't explain how Earth and the moon have similar rocks.

In 1969, astronauts from the United States first visited the moon to study it. Twelve people have walked on its surface since then. NASA's new Artemis program seeks to send people to the moon for the first time since 1972.

## Consider This

Scientists know the moon is made of rocks similar to those on Earth. How do you think the moon was formed?

# Mars' Two Moons

Mars is the closest planet to Earth that also has moons. The fourth planet from the sun has two known moons, Phobos and Deimos. Astronomer Asaph Hall discovered them in 1877. Both moons are small and have low gravity. Phobos is about 16.5 miles (26.6 km) long at its longest point. Deimos is about 9 miles (15 km) long at its longest point. They are both shaped like potatoes.

Phobos, seen here in the middle, orbits Mars closer than any moon orbits a planet in the solar system.

Stickney Crater, seen at the right of this image, was named after mathematician Chloe Angeline Stickney Hall. She was married to Asaph Hall.

Phobos is very close to Mars. It is about 5,825 miles (9,375 km) from the planet's center. Mars's gravity is so strong that Phobos will someday crash into the planet's surface. One of Phobos' craters, named Stickney, is about half as wide as Phobos itself. This giant crater came from an impact that nearly shattered the entire moon in the past.

## Consider This

Phobos and Deimos are very small. Why do you think Mars' moons are not round?

# Jovian Moons

Jupiter is the largest planet in our solar system and is the fifth planet from the sun. It is mostly made of gases, and it is the first of the planets called "gas giants." It has between 80 and 95 known moons—together known as the Jovian moons. Astronomers on Earth using large telescopes have been searching for more. In 1610, Jupiter's four brightest moons were the first space objects discovered by astronomer Galileo Galilei with a telescope. Those moons were named Ganymede, Io, Europa, and Callisto.

Scientists believe Ganymede may have more underground saltwater than all of the water on Earth.

NASA's Galileo spacecraft captured this image showing a volcano erupting on Io.

Ganymede is the largest moon in the solar system. It is larger than the planet Mercury. Its iron core creates its own magnetic field, the only known moon to do so. Its surface is very icy and may have water under it. Io has gigantic, active volcanoes on its surface. It has more volcanic activity than any planet or moon in the solar system thanks to the gravitational pull on it from Jupiter as well as fellow large moons Ganymede and Europa.

## Consider This

Many rocks form an asteroid belt between the orbits of Mars and Jupiter. Why don't scientists call these rocks moons?

# Saturn and Its Many Moons

Saturn, the sixth planet from the sun, has more moons than any other planet in the solar system. The gas giant has 274 confirmed moons! Titan is Saturn's largest moon and the second largest in the solar system. Discovered in 1655, it was the first moon found after Jupiter's first four moons were discovered. It is the only moon that has clouds. Its atmosphere is made of thick gases such as nitrogen and methane. It also has lakes of ethane on its surface.

Saturn's rings are made of ice and rock but are not moons. However, many small moons help keep the rings in place. The moon Pan was discovered in 1990 between two of the rings. Other moons like Pan keep the edges of nearby rings sharp and distinct.

Scientists discovered a whopping 128 small moons orbiting Saturn in 2025, nearly doubling its count!

The atmosphere on Titan gives it this hazy appearance.

## compare and contrast

People have landed a spacecraft on Titan. Compared to Earth's moon, what is different about Titan?

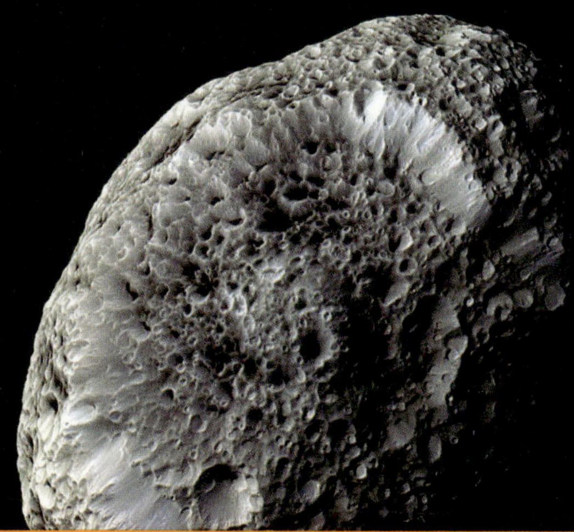

The irregular-shaped Hyperion has many deep impact craters. Scientists believe this is because the moon has low density.

# Uranus and the Literary Moons

Uranus, the seventh planet from the sun, has 28 known moons, all named after literary characters. Its five major moons are Oberon, Titania, Ariel, Umbriel, and Miranda. Uranus' other moons are very small. Eight of the small moons are so close together that scientists are not sure how they avoid crashing into one another. Scientists think all of Uranus' moons are made of frozen water and rock.

Miranda is the smallest of the major moons. Its surface is very strange. It has some spots with huge craters. Other areas have big valleys and cliffs. It looks like many pieces of rock and ice were smashed together. Miranda may have been broken apart by crashing into other objects, then put back together again. This moon has giant canyons up to 12 times as deep as the Grand Canyon.

Several of the largest moons orbiting Uranus (center with rings) can be seen in this photo taken by a telescope in Chile.

## compare and contrast

Compare and contrast the shape of Miranda and Earth's Moon. What is different about the two moons?

Miranda is named after one of the characters in William Shakespeare's play *The Tempest*, which is another name for a storm.

# Neptune and Triton

The farthest planet from the sun, Neptune, has 16 known moons. Its largest is named Triton. Triton was discovered in 1846, about a month after scientists found Neptune. Triton is nearly as big as Earth's moon.

Scientists believe that Triton might have formed as an independent planet. Because of gravity, Neptune may have pulled Triton into its orbit. Unlike all the other large moons in the solar system, Triton revolves around the planet in a direction opposite to Neptune's spin.

Triton has the coldest surface discovered anywhere in our solar system, about -391° F (-235° C).

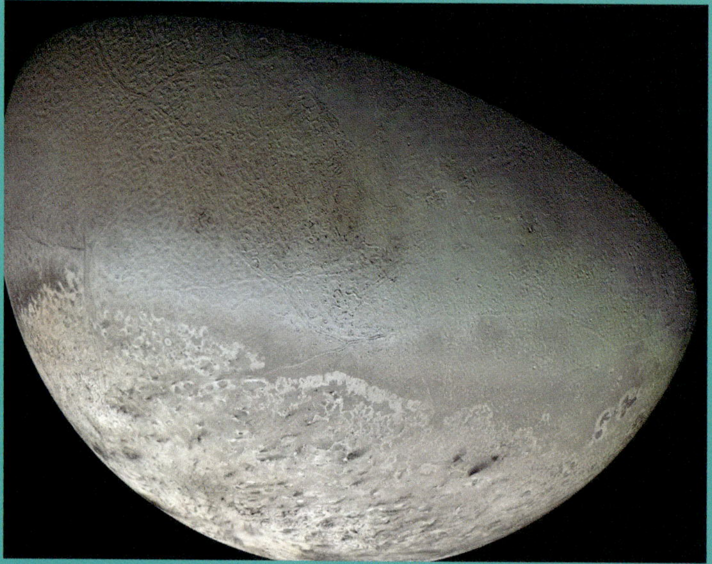

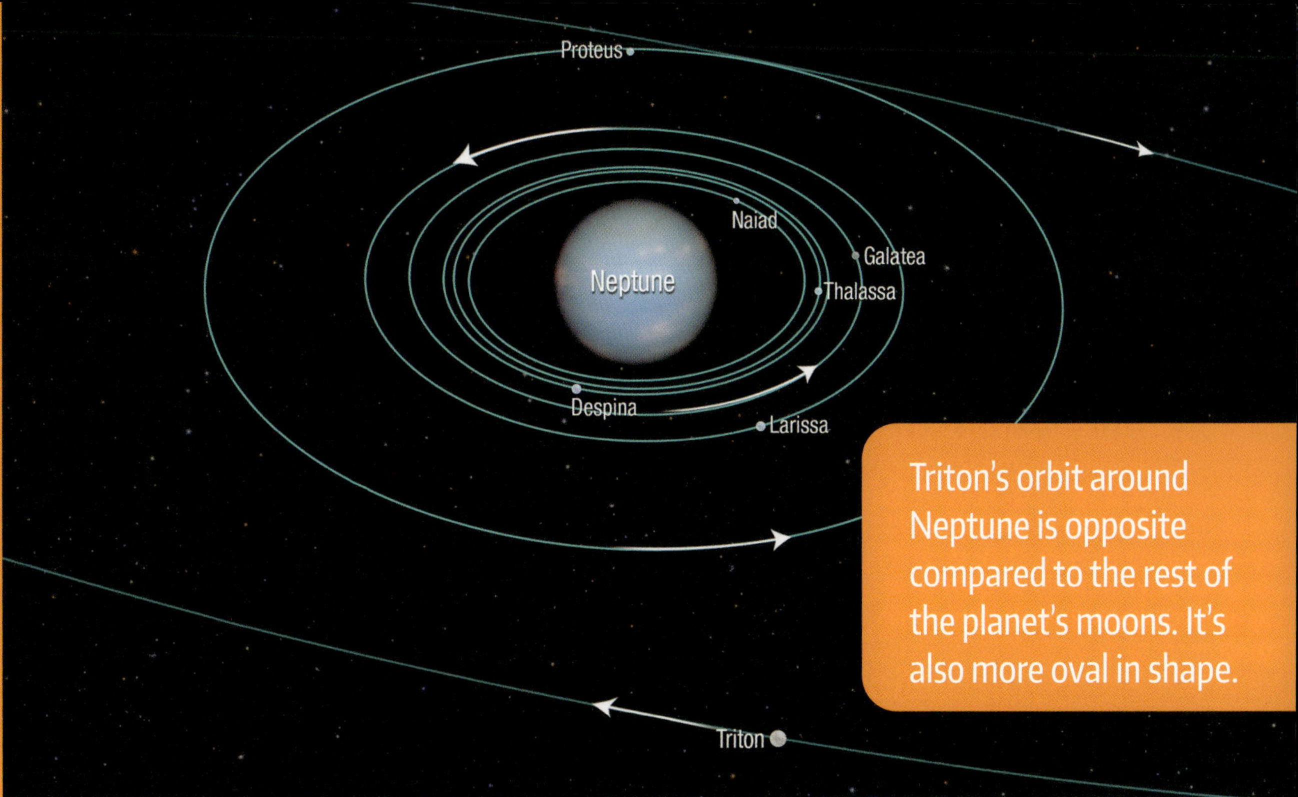


Triton's orbit around Neptune is opposite compared to the rest of the planet's moons. It's also more oval in shape.

Nereid was discovered in 1949. Its orbit is the most oval-shaped of all the known moons in the solar system. Scientists believe Nereid is a captured asteroid. Its orbit also could have been changed when Triton was captured by Neptune's gravity.

Neptune is named after the Roman god of the sea, and its moons are named after other gods and figures from mythology.

## WORD WISE

**THE SHAPE OF A MOON'S ORBIT IS CALLED ITS ECCENTRICITY. ITS MEASUREMENT IS DETERMINED BY HOW MUCH THE ORBIT DIFFERS FROM A PERFECT CIRCLE.**

# Pluto Has Moons!

Pluto was once the ninth planet in our solar system, but in 2006, scientists downgraded it to a **dwarf planet**. It now belongs to a group of rocks and asteroids at the edge of the solar system called the Kuiper Belt.

Pluto has five moons. Charon, with a diameter of nearly 750 miles (1,210 km), was discovered in 1978. Nix and Hydra followed in 2005. A fourth moon, Kerberos, was found in 2011. A fifth moon was named Styx in 2013.

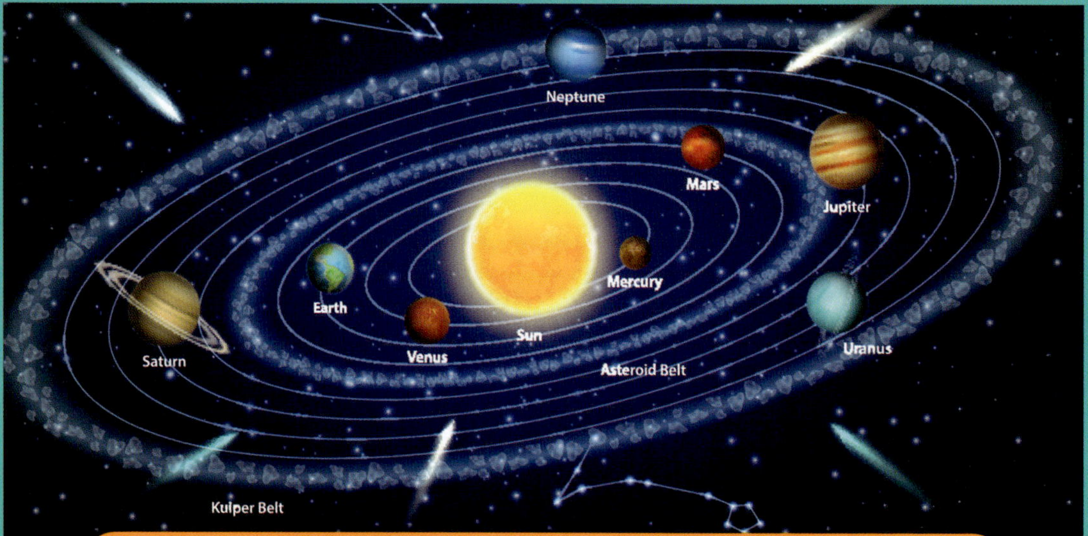


Scientists believe the Kuiper Belt is home to millions of small, icy objects, including some that may be as big as Pluto.

In addition to being close in size, both Pluto (right) and Charon have similar areas of red terrain on their surfaces.

Pluto and Charon are closer in size than any of the planets and their moons. Scientists are always discovering new bodies in the Kuiper Belt. Although none is considered a planet, many of them have moons. Scientists believe the objects in this region are leftovers from the early days of our solar system's formation.

## WORD WISE

**A DWARF PLANET IS A WORLD THAT ORBITS THE SUN BUT ISN'T PERFECTLY ROUND AND HASN'T BEEN ABLE TO CLEAR ITS ORBITAL PATH OF RUBBLE.**

# Asteroids with Moons

Planets and dwarf planets aren't the only objects in the night sky to have moons. Scientists have even found asteroids with their own moons. In 1884, astronomer Johann Palisa discovered an asteroid that he named Ida. The Galileo spacecraft passed Ida in 1993 and took pictures of it. Scientists noticed a small rock orbiting Ida. It was the first moon discovered that orbited something other than a planet. They named the moon Dactyl, after a group of beings from Greek mythology who lived on Mount Ida.

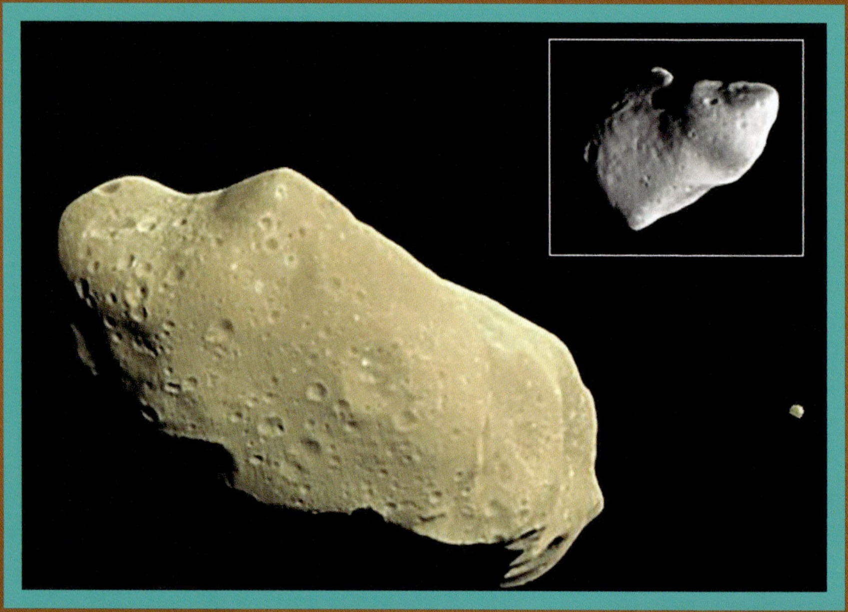

The asteroid Ida (left) has a much smaller moon named Dactyl, a close-up of which can be seen in the upper box.

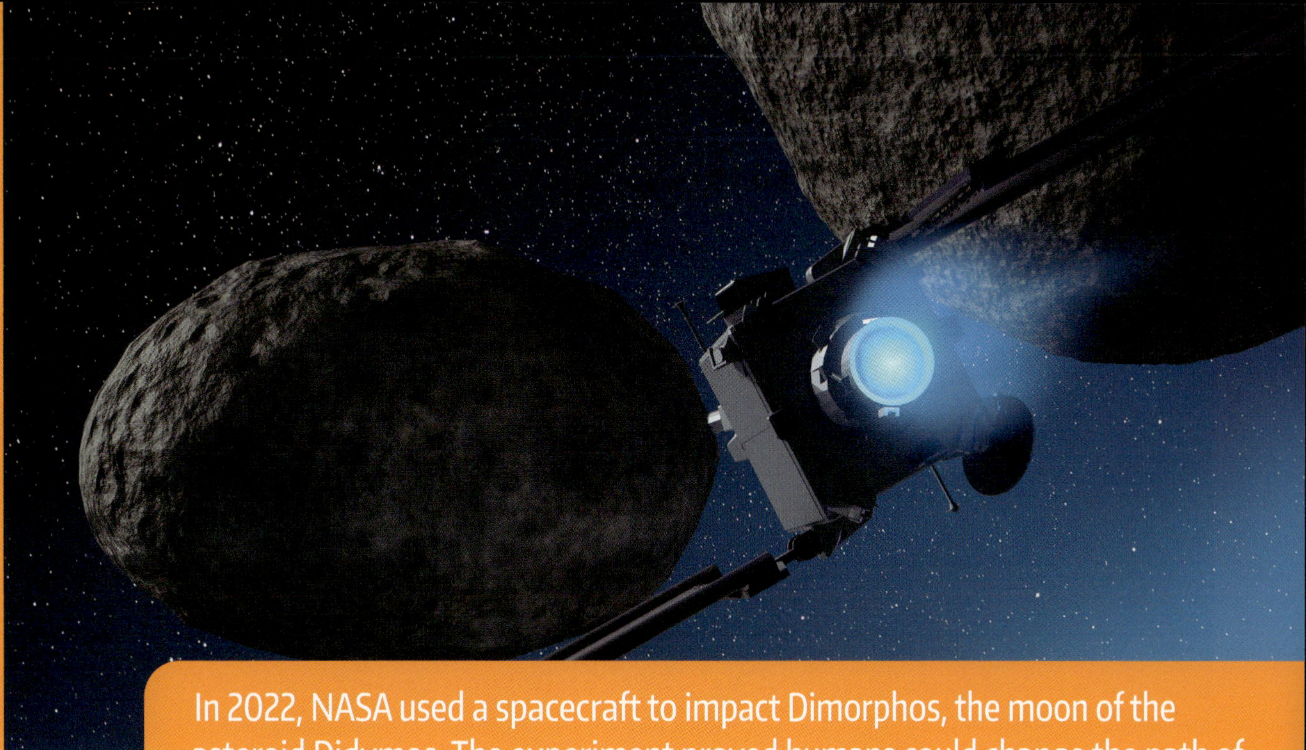

In 2022, NASA used a spacecraft to impact Dimorphos, the moon of the asteroid Didymos. The experiment proved humans could change the path of an object's orbit and one day protect Earth from an asteroid impact.

A tiny speck in photographs, Dactyl is 1 mile (1.6 km) at its widest point. It orbits about 56 miles (90 km) away from Ida. Scientists have discovered about 470 other moons orbiting dwarf planets, asteroids, and bodies called trans-Neptunian objects. Trans-Neptunian objects are bodies that exist beyond Neptune's orbit.

## Consider This

Moons can orbit anything that has its own gravity. Do you think a moon could have its own moon?

# New Discoveries

Today, scientists know of 891 moons in our solar system. Humans once called the Earth's moon a planet but now we know moons are bodies that orbit planets and objects other than stars. Some moons are just rocks, while others are made of ice. One moon, Titan, has its own atmosphere. As humans explore the solar system, they find other moons and learn more about them.

Finding moons outside of our solar system might give scientists new clues about how they form.

NASA launched its powerful James Webb Space Telescope in 2021. Among other goals, scientists plan to use it in their search for exomoons.

The definition of what a moon can be is always changing as scientists learn new things. As people explore deeper into space, they find more moons orbiting planets. They may even find moons outside the solar system, called exomoons. In fact, scientists think there may be moons around exoplanets Kepler-1625b and Kepler-1708 b, thousands of light years away from Earth.

## Consider This

Humans have not walked on Earth's moon since 1972. Do you think people should go back to visit the moon?

# Glossary

**asteroid belt** A group of asteroids orbiting the sun between Mars and Jupiter.

**astronomer** A scientist that studies space.

**atmosphere** A mass of gases surrounding a body.

**density** How much mass is in a certain volume.

**international** Involving two or more countries.

**meteorite** A piece of rock or metal that has fallen from space to the ground.

**mythology** Stories told in ancient culture to explain a practice, belief, or natural happening.

**orbit** The path taken by a body circling around another body.

**reflect** To give back light.

**revolve** To move around another object; to orbit.

**rotation** One complete turn.

**telescope** A tool that makes faraway objects look bigger and closer.

**temperature** A measurement that indicates how hot or cold something is.

**volcano** An opening in a planet's or moon's surface through which hot liquid rock or ice sometimes flows.

# For More Information

## Books

Labrecque, Ellen. *Mysteries of Moons and Moon Phases.* North Mankato, MN: Capstone Press, 2021.

Loy, Harriet. *Moons.* Minneapolis, MN: Bellwether Media, 2023.

Miller, Ron. *Natural Satellites: The Book of Moons.* Minneapolis, MN: Twenty-First Century Books, 2021.

## Websites

**NASA: Moons of Our Solar System**

*www.science.nasa.gov/solar-system/moons*

Visit the official NASA website to read more about the known moons in our solar system.

**National Geographic Kids: The Phases of the Moon**

*www.natgeokids.com/uk/discover/science/space/the-phases-of-the-moon*

Read more about the phases of Earth's moon and why it appears to change shape each night.

Publisher's note to educators and parents: Our editors have carefully reviewed these websites to ensure that they are suitable for students. Many websites change frequently, however, and we cannot guarantee that a site's future contents will continue to meet our high standards of quality and educational value. Be advised that students should be closely supervised whenever they access the internet.

# Index